George Y. Browne

Browne's Arithmetical Tables

Combined with easy lessons in mental arithmetic for beginners

George Y. Browne

Browne's Arithmetical Tables
Combined with easy lessons in mental arithmetic for beginners

ISBN/EAN: 9783337393281

Printed in Europe, USA, Canada, Australia, Japan

Cover: Foto ©berggeist007 / pixelio.de

More available books at **www.hansebooks.com**

BROWNE'S

Arithmetical Tables,

COMBINED WITH

EASY LESSONS

IN

MENTAL ARITHMETIC.

FOR BEGINNERS.

BY REV. GEO. Y. BROWNE, A. M.

TUSCALOOSA, ALA.

ATLANTA, GA.:

FRANKLIN PRINTING HOUSE.

J. J. TOON, PUBLISHER AND PROPRIETOR.

1865.

PREFACE.

No apology is deemed necessary for presenting this little work to the public. As far as it goes, it contains the results of more than twenty years' experience in practical instruction in this department of the duties of the Southern school room.

One important feature, too frequently overlooked by those who prepare books of instruction for very little children, has been steadily kept in view in its preparation. The book is adapted to the supposed progress in reading of its little students. Indeed, it is thought that it may be used profitably as a reading book for the class of pupils for whom it is designed.

Another feature is that the thinking faculty of the learner is, from the beginning, kept in constant, but not too laborious exercise. It is hoped that the lessons are sufficiently progressive for the brightest pupils, yet not too much so for those of average ability.

Whether the series, of which this may perhaps be the "Primer," will ever be completed, will depend partly upon the reception it may meet with, and partly upon the author's other pressing duties in the school room and in the pulpit.

G. Y. B.

Eufaula, Alabama.

INTRODUCTION.

The grand object, never to be lost sight of, in every department of instruction, should be the development of the intellectual faculties. Mathematical studies have generally been thought to be highly conducive to this grand object; but, as ordinarily pursued, it may well be questioned whether they do not exert an injurious rather than a beneficial influence. The author hopes, by this "Primer," to aid both teacher and pupil to draw the highest possible intellectual benefit from these elementary studies.

MACHINERY.—In the prosecution of this labor but little machinery is necessary. A black board and some chalk are great conveniences. An Abacus may serve for a while to amuse, and may prove an assistance. The author had one for some time, but found it rather an incumbrance. A slate and pencil from the beginning will be found indispensable.

RECITATIONS.—Three features should characterize every recitation, viz: absolute perfectness, great promptness, and rigid attention. To state these points may be sufficient. The author has been accustomed to consider a slip of the tongue, or the

slightest hesitation in reciting the tables, a failure.
It requires a great deal of drilling to stimulate pupils to this very high standard of excellence, but it
may be accomplished by gradual approaches, not
only without annoyance, but with positive pleasure
to the pupil.

THE FIGURES.—To teach the forms and names of
the figures, make the digits on the black-board in
any order whatever, thus:

$$1 \qquad 9 \qquad 4 \qquad 0$$
$$5 \qquad 3 \qquad 6 \qquad 5$$
$$7 \qquad 2 \qquad 8 \qquad 4$$

Name them over in concert with the class several
times, and then call upon each pupil to point to any
number that may be named. A few lessons of from
ten to fifteen minutes each will be sufficient for the
instruction of a large class. The pupils should be
encouraged to write the figures with chalk upon the
black-board. The value of the figures may be
taught by making marks upon the black-board and
writing in connection the proper figures, thus: 1=1
11=2 111=3 1111=4 11111 1111=9, and so on.
The pupils should, of course, write such exercises
both upon the board and upon the slates.

The same point may be further illustrated by an

Abacus, or by buttons, marbles, pencils, books or anything else that may be convenient.

COUNTING.—This exercise should be performed forwards and backwards to any required number in concert at first, but afterwards by each pupil singly. By a simple device, which will be found explained at some length in the remarks before Multiplication, the pupil will be enabled to count by any intervals that may be required. It may be done in the first place by the odd numbers, 1, 3, 5, 7 and 9, and afterwards by the even numbers, 2, 4, 6 and 8. As this kind of counting facilitates the acquirement of all the four fundamental operations of arithmetic, it is suggested that it be put into practice as soon as the pupils have a tolerable idea of the value of the figures, and kept in daily exercise, forwards and backwards, until the whole school can count thus promptly and accurately.

THE SIGNS.—The author prefers to use Algebraic signs from the beginning. In the *Addition table* two signs are used +, plus; and =, equal. The *name* of the first is plus—its meaning is "more, add or and." In the table, the last meaning is used. The = is the sign of equality, and may mean "are, leaves or is," as the case may require. Thus, in the Addition table for $1+1=2$ read one, and one are two, and so on.

In the *Subtraction Table* one additional sign is

used, viz : —, minus. It means less or from. Hence
read (5 — 1 = 4) thus : 1 from 5 leaves 4. &c.

In the Multiplication Table, another sign is em-
ployed. It is formed thus ×. and named the sign
of multiplication. It may be interpreted "times."
The third line of the table of six, which is written
thus (3 × 6 = 18.) may be read three times six is
eighteen. (*See foot note Multiplication.*)

In Division, still another sign, made thus ÷, is
used. It is called the sign of division, and may be
interpreted by "into." Thus in the table of nine
the sixth line, which is written (54 ÷ 9 = 6,) may
be read nine into fifty-four six times, and so on.

ADDITION.

Lesson First.—-Increasing by Ones.

1. One boy is on a box, one more boy gets on; how many boys are now on the box?

2. Two hens are in a coop, one more is put in; how many hens are now in the coop?

3. Three nuts are on a plate; if one nut more is put on, how many will that make?

4. John has four balls and James has one; how many balls have both boys?

5. Five girls are on a bench and one is on a chair; how many girls are there in all?

6. Six ducks are in the pond; if one more goes in how many will that make?

7. I have seven dimes in my right hand and one in my left; how many dimes are in both hands?

8. Add eight yards to one yard; how many yards will there be?

9. Nine sticks of wood are in a pile; if one more stick is put in the pile, how many will that make?

10. Ten hats are in a box; put one more hat in, and how many hats will there be?

THE TABLE.

1	+ 1	= 2		6	+ 1	= 7
2	1	3		7	1	8
3	1	4		8	1	9
4	1	5		9	1	10
5	1	6		10	1	11

EXERCISES.

5 and 1? 3 and 1? 7 and 1? 2 and 1? 3 and 1?
8 and 1? 4 and 1? 2 and 1? 1 and 1? 5 and 1?
2 and 1? 7 and 1? 3 and 1? 9 and 1? 10 and 1?

Lesson Second.—Increasing by Twos.

1. Mary has put two pins in her dress and wants one more; how many will that make?

2. One hen eats two grains of corn and then eats two more grains; how many grains did she eat?

3. John puts two chairs to the table and Jane puts three; how many are there at the table?

4. If four books are in a pile, and two more are put in, how many books will there be in the pile?

5. George drives five nails into a board, and two into another; how many nails did he drive?

6. There are six boxes on a dray, and two more are put on; how many does that make?

7. Seven bells are in a store, and two more are brought in; how many does that make?

8. Eight hats are on a stand; if two more are put on how many will there be?

9. James has nine old coats and two new ones; how many coats has he?

10. Ten black birds and two red ones make how many?

THE TABLE.

1	+	2	=	3		
2		2		4		
3		2		5		
4		2		6		
5		2		7		

6	+	2	=	8		
7		2		9		
8		2		10		
9		2		11		
10		2		12		

EXERCISES.

4 and 2?	5 and 2?	4 and 1?	3 and 2?	7 and 2?
6 and 2?	7 and 1?	10 and 2?	1 and 2?	4 and 1?
7 and 2?	9 and 2?	8 and 2?	6 and 1?	2 and 2?
3 and 1?	2 and 2?	6 and 2?	5 and 2?	4 and 2?
9 and 1?	8 and 2?	2 and 2?	5 and 1?	6 and 2?
7 and 2?	5 and 2?	9 and 2?	4 and 2?	5 and 2?

☞ Many more such questions should be asked.

Lesson Third.—Increasing by Threes.

1. Three caps are in the drawer, and one more is put in ; how many will that make ?

2. Jane makes up three beds in one room and two in another ; how many beds does she make up ?

3. Three large figs and three small ones make how many figs ?

4. If four jugs are full of milk and three full of wine, how many jugs will that make ?

5. Five bags are filled with nuts and three with rice ; how many bags does that make ?

6. There are six kegs in a row and three in another ; how many kegs in both rows ?

7. Seven axes with helves and three without, make how many axes ?

8. Mary has eight dresses and Jane has three; how many dresses have both ?

9. Nine men are in the house and three in the street ; how many men does that make ?

10. There are ten maps in the book and three on the wall ; how many maps are there in all ?

THE TABLE.

1	+ 3	= 4		6	+ 3	= 9
2	3	5		7	3	10
3	3	6		8	3	11
4	3	7		9	3	12
5	3	8		10	3	13

EXERCISES.

7 and 3?	9 and 2?	5 and 3?	6 and 2?	8 and 3?
3 and 3?	6 and 1?	7 and 2?	3 and 3?	5 and 2?
5 and 2?	2 and 3?	9 and 3?	1 and 3?	2 and 3?
4 and 1?	7 and 2?	8 and 3?	7 and 1?	6 and 3?
4 and 2?	5 and 1?	3 and 1?	2 and 5?	9 and 3?
7 and 3?	5 and 2?	2 and 1?	4 and 1?	7 and 2?

Lesson Fourth.—Increasing by Fours.

1. Four palm-leaf fans and one silk one will make how many fans?

2. John brings in four logs and James brings in two; how many logs do both bring in?

3. There are four glass mugs and three tin ones; how many mugs does both make?

4. Mary has four white and four red hair pins; how many hair pins has she?

5. Jane found five eggs in one nest and four in another; how many eggs did she find?

6. Sallie has four dolls and Mary has six; how many dolls have both?

7. George drew seven cats on one side of his slate and four on the other; how many cats did he draw?

8. Eight birds are on one tree and four on another; how many birds on both trees?

9. A boy ate nine cakes and a girl ate four; how many did both eat?

10. Ten black eyes in five girls and four blue eyes in two girls make how many eyes?

THE TABLE.

1	+ 4	= 5	6	+ 4	= 10
2	4	6	7	4	11
3	4	7	8	4	12
4	4	8	9	4	13
5	4	9	10	4	14

EXERCISES.

9 and 4?	7 and 4?	3 and 3?	6 and 2?	8 and 4?
3 and 4?	4 and 4?	6 and 4?	4 and 3?	3 and 4?
5 and 3?	2 and 2?	9 and 1?	1 and 4?	7 and 2?
7 and 2?	5 and 3?	8 and 4?	7 and 4?	4 and 3?
5 and 4?	2 and 4?	2 and 1?	7 and 3?	9 and 3?
6 and 4?	7 and 4?	.9 and 4?	2 and 4?	7 and 1?

Lesson Fifth.—Increasing by Fives.

1. A girl has five toys; if one more be given her how many toys will she have?

2. There are two balls of white yarn in a box and five of red; how many balls are there?

3. Five mules are in one team and three in another; how many mules in both teams?

4. Mary has five pet lambs and Kate has four; how many lambs have both?

5. Five square bales and five round bales make how many bales?

6. Six pinks are in bloom and six are not; how many pinks does that make?

7. Seven sheets are in the wash and six on the beds; how many sheets in all?

8. John jumps five times, Pete eight times; how many times do both jump?

9. Nine guns in one rack and five in another make how many guns?

10. James gave ten cents to a poor man and Charles gave five; how many cents did both give?

THE TABLE.

1	+ 5	= 6		6	+ 5	= 11
2	5	7		7	5	12
3	5	8		8	5	13
4	5	9		9	5	14
5	5	10		10	5	15

EXERCISES.

7 and 5?	9 and 1?	5 and 5?	6 and 5?	8 and 5?
3 and 4?	6 and 5?	7 and 5?	3 and 5?	5 and 4?
4 and 5?	2 and 5?	9 and 4?	1 and 5?	2 and 3?
5 and 3?	4 and 3?	3 and 4?	4 and 5?	5 and 4?
5 and 5?	7 and 5?	3 and 1?	9 and 5?	7 and 5?
2 and 5?	5 and 5?	9 and 5?	10 and 5?	10 and 4?

Lesson Sixth.—Increasing by Sixes.

1. The babe had six teeth and one more has come; how many teeth has it now?

2. A boy had six fish hooks, he found two more; how many did that make?

3. If three pies are done and six are not done; how many pies will that be?

4. Four trees are in one, row and six in another; how many trees in both rows?

5. One knife has five blades, another has six; how many blades have both knives?

6. There are six doors in the old house and six in the new; how many doors are there in both houses?

7. If seven mills are on one stream and six on another; how many will there be on both streams?

8. Eight men are at work on that side of the house and six on this; now many men are there on both sides?

9. If there are nine hens that have nests and six that have not; how many hens will that make?

10. There are ten spools of white thread and six of black; how many spools of both?

THE TABLE.

1	+	6	=	7	6 + 6 = 12
2		6		8	7 6 13
3		6		9	8 6 14
4		6		10	9 6 15 ·
5		6		11	10 6 16

EXERCISES.

5 and 6? 3 and 6? 9 and 5? 6 and 4? 2 and 3?
4 and 6? 7 and 6? 5 and 4? 9 and 6? 6 and 5?
3 and 4? 1 and 6? 4 and 6? 6 and 3? 9 and 2?
7 and 5? 3 and 6? 8 and 6? 6 and 6? 9 and 5?
2 and 6? 5 and 4? 4 and 6? 7 and 3? 2 and 4?
3 and 3? 7 and 5? 10 and 3? 10 and 5? 10 and 6?

Lesson Seventh.—Increasing by Sevens.

1. One loaf of bread is at the fire and seven are on plates; how many loaves does that make?
2. Two cars and seven cars will make a train of how many cars?
3. John has three balls, James gives him seven more; how many has he then?
4. Seven pine posts are on this side of the fence and four on that; how many pine posts are there?
5. Five plows and seven plows make how many?
6. Six boys have hoes and seven have guns; how many boys does that make?
7. Ann has seven pins in one hand and seven in the other; how many are there in both hands?
8. Eight houses are on one side of the road and seven on the other, make how many houses?
9. Nine boys are in one room and seven in another; how many boys are there in both rooms?
10. Ten cows and seven calves make how many cattle?

THE TABLE.

1	+ 7	= 8		6	+ 7	= 13
2	7	9		7	7	14
3	7	10		8	7	15
4	7	11		9	7	16
5	7	12		10	7	17

EXERCISES.

6 and 7? 5 and 7? 10 and 4? 4 and 5? 9 and 7?
5 and 7? 3 and 7? 6 and 6? 5 and 4? 2 and 5?
3 and 7? 1 and 3? 4 and 7? 6 and 5? 7 and 4?
9 and 7? 4 and 3? 10 and 6? 8 and 1? 6 and 4?
2 and 7? 5 and 1? 10 and 1? 4 and 2? 3 and 5?
4 and 6? 1 and 1? 9 and 4? 2 and 9? 1 and 8?
7 and 6? 8 and 3? 10 and 9? 9 and 2? 7 and 8?
7 and 5? 1 and 9? 6 and 1? 9 and 9? 10 and 7?

Lesson Eighth.—Increasing by Eights.

1. Eight leaves have blots on them and one leaf is torn; how many leaves does that make?
2. Two trees are blown down and eight are dead; how many trees?.
3. Three sheep are in the fold and eight more are going in; how many will there be in?
4. James has four books in his desk and John has eight in his; how many are there in both desks?
5. Five snakes are under one log and eight under another; how many snakes under both logs?
6. Three men have six spurs and four boys have eight; how many spurs do they all have?
7. There are seven blots on one page and eight on another; how many blots are there on both pages?
8. There are eight wheels to this car and eight wheels to that; how many wheels to both cars?
9. If one boy makes eight rings on a slate and another boy makes nine, how many will both make?
10. Ten wet days and eight dry days; make how many?

THE TABLE.

1	+ 8 =	9	6	+ 8 =	14
2	8	10	7	8	15
3	8	11	8	8	16
4	8	12	9	8	17
5	8	13	10	8	18

EXERCISES.

8 and 5?	4 and 8?	3 and 5?	7 and 4?	9 and 6?
5 and 7?	6 and 8?	8 and 8?	6 and 8?	7 and 4?
3 and 7?	5 and 6?	2 and 8?	9 and 4?	1 and 8?
8 and 5?	10 and 8?	9 and 8?	3 and 7?	7 and 8?
1 and 6?	10 and 7?	7 and 5?	5 and 8?	3 and 8?
5 and 7?	9 and 4?	3 and 2?	7 and 6?	1 and 6?
9 and 2?	10 and 5?	7 and 7?	8 and 4?	3 and 6?

Lesson Ninth—Increasing by Nines.

1. Mary knows nine tunes; if she learns another how many will she know?

2. Two boys are at the well and nine are playing town-ball; how many boys are there?

3. Three desks are of oak wood and nine of pine; how many desks in all?

4. Four goobers (ground-nuts) are on the stove; if nine more are put on how many will that make?

5. Five frogs are in the well, but nine have been taken out; how many were there in the well?

6. How many corks do six large ones and nine small ones make?

7. There are seven rods for the girls and nine for the boys; how many rods were there in all?

8. If to eight pounds of rice you add nine pounds, how many pounds will there be?

9. How many capes has Mary if she has nine of cloth and nine of lace?

10. Ten rows of corn are on one ear and nine on another; how many rows are on both ears?

THE TABLE.

1 + 9 = 10			6 + 9 = 15		
2	9	11	7	9	16
3	9	12	8	9	17
4	9	13	9	9	18
5	9	14	10	9	19

EXERCISES.

4 and 9?	10 and 8?	3 and 9?	6 and 9?	9 and 5?
7 and 9?	6 and 3?	4 and 7?	7 and 9?	5 and 6?
9 and 9?	6 and 4?	4 and 8?	10 and 9?	8 and 9?
2 and 1?	6 and 9?	8 and 9?	9 and 6?	7 and 5?
1 and 9?	9 and 4?	2 and 9?	5 and 1?	7 and 8?
5 and 9?	6 and 5?	9 and 4?	7 and 2?	4 and 9?

Lesson Tenth.—Increasing by Tens.

1. One man owns one goat and another owns ten; how many do both own?

2. John brings two yams to school and Mary brings ten; how many do both bring?

3. Three white beads and ten green beads make how many?

4. Four Marys and ten Anns make how many girls?

5. How many keys are there if five are in one bunch and ten in another?

6. There are six shell combs and ten horn combs; how many combs in all?

7. If one girl is seven years old and another is ten, what is the sum of their ages?

8. Eight gnats are on my face and ten are on the wing; how many gnats are there?

9. If there are nine red calves and ten black ones, how many calves are there in all?

10. Ten birds are in one pie and ten in another; how many birds are there in both pies?

THE TABLE.

1	+ 10	= 11		6	+ 10	= 16
2	10	12		7	10	17
3	10	13		8	10	18
4	10	14		9	10	19
5	10	15		10	10	20

EXERCISES.

1 and 10? 7 and 9? 5 and 4? 3 and 10? 5 and 6?
2 and 10? 7 and 4? 10 and 9? 8 and 10? 7 and 6?
3 and 10? 9 and 7? 10 and 8? 4 and 10? 9 and 6?
6 and 10? 5 and 4? 10 and 4? 6 and 10? 8 and 5?
9 and 10? 3 and 7? 7 and 4? 2 and 10? 8 and 5?
5 and 10? 5 and 8? 4 and 9? 9 and 7? 2 and 9?
9 and 6? 4 and 7? 5 and 9? 8 and 7? 8 and 5?

Lesson Eleventh.—Miscellaneous Questions.

1. Mary has one apple and George gives her three more; how many has she then ?

2. In a box there are six marbles; John puts in four more ; how many are there in the box ?

3. Seven cherries are on one stick and nine on another; how many cherries are there on both sticks ?

4. Eight birds are sitting on one limb of a tree, and nine more are on another limb; how many on both limbs ?

5. Nine houses are on this side the street and seven on that; how many houses are there on both sides ?

6. The speckled hen has eleven chickens and the black hen seven; how many have both ?

7. John has learned twelve verses and little Kittie has learned five; how many have they both learned ?

8. On one dish there are fourteen eggs and on another six; how many on both dishes ?

9. Sallie has four kittens and Mary has three dolls ; how many have both ? (If answered, ask "seven what?" and state that things which are not alike cannot be added.)

10. Five boys and three boys and four boys; are how many ?

11. Three kites, two kites, and five kites; are how many ?

12. Six pens, three pens, and two pens ; are how many ?

13. One horse, six horses, and five horses; are how many horses ?

14. Three toes, five toes, and seven fingers; are how many ? Why ?

15. Seven books, five books, and four books; are how many ?

16. Nine gloves, four gloves, and seven gloves ; are how many ?

17. Eight chairs, six chairs, and nine chairs; are how many ?

18. Four birds, nine birds, and five birds ; are how many?

19. Six stars, eight stars; and nine stars ; are how many ?

20. Nine beds, three beds, two beds, and one bed ; are how many ?

EXERCISES.

4+6+2?	9+2+3?	3+6+2?	3+4+6+2?
5 2 3?	2 7 2?	6 4 3?	5 3 2 4?
7 2 4?	3 6 4?	7 6 2?	8 4 3 1?
3 7 5?	5 4 2?	6 5 3?	6 3 5 2?
1 8 2?	6 3 5?	4 3 6?	5 4 3 2?
5 4 3?	8 5 4?	5 4 8?	3 7 2 5?
3 2 6?	2 3 6?	6 3 2?	7 6 3 4?
6 3 4?	9 1 3?	4 5 3?	8 4 5 3?
5 2 3?	8 4 2?	5 6 2?	5 3 8 4?
4 2 5?	7 5 6?	7 2 4?	9 5 4 2?
3 4 6?	6 4 3?	6 3 5?	8 3 6 4?
7 3 2?	5 3 2?	3 2 5?	5 4 7 6?

NOTE.—The Teacher should ask other questions of the same kind as are written on these two last pages. Require the answer to be given immediately upon the completion of the question. The object of this is to train the faculty of attention, and to compel the pupil to use his head rather than his fingers in the calculation. Let me illustrate this: Suppose the teacher says, (and he should always speak somewhat slowly) "nine, five" the pupil should be accustomed to *think* fourteen; then should the teacher say, "and three," the pupil is ready to answer seventeen.—The author has sometimes had quite small children so well trained that they would tell accurately and promptly upon the completion of the question, the sum of as many as a dozen figures thus given out.

SUBTRACTION.

Lesson Twelfth.—Decreasing by Ones.

1. A boy had one pencil, but he soon lost it; what did he have left?

2. Mary had two yards of silk; she cut off one yard for a dress for her doll; how much remained?

3. Three little boys stood up to recite, but one was sent to his seat for talking; how many remained?

4. Four marbles were in a ring, but John knocked out one; how many remained?

5. Sarah, having five apples, ate one; how many remained?

6. A planter had six cows, one of which he sold; how many were left.

7. Take one feather from a bunch of seven feathers; how many will remain?

8. Eight birds were on a branch, but one flew away; how many were left?

9. A hen had nine little chickens, but the cat ate one of them; how many were left?

10. Ten books were in a pile, but George takes one off; how many are left?

THE TABLE.

1	— 1	= 0		6	— 1	= 5
2	1	1		7	1	6
3	1	2		8	1	7
4	1	3		9	1	8
5	1	4		10	1	9

EXERCISES.

5 less 1?	7 less 1?	6 less 1?	4 less 1?
9 less 1?	6 less 2?	5 and 3?	7 less 1?
5 less 3?	8 less 1?	8 and 6?	9 less 8?
10 less 1?	6 less 1?	10 less 6?	3 less 1?

22

ARITHMETIC.

Lesson Thirteenth.—Decreasing by Twos.

1. Two vials were standing on a shelf, but both were knocked down; what remained?

2. Three tapers were lighted, but the wind blew two of them out; how many continued to burn?

3. Four giants went out to war; two of them were killed; how many returned?

4. Five basins were packed in a box; two were broken; how many were whole?

5. Six spiders were in a corner; Jane swept two of them down; how many were left?

6. Seven papers are lying together, but two of them are torn; how many are whole?

7. Eight pies were on the safe; two of them were eaten at dinner; how many are left?

8. Mary has nine cherries, two of which are wormy; how many are sound?

9. Ten cows are in the field, but two of them are without horns; how many have horns?

10. Eleven trees were standing in a row, but two have been cut down; how many remain?

THE TABLE.

2	—	2	= 0	7	—	2	= 5
3		2	1	8		2	6
4		2	2	9		2	7
5		2	3	10		2	8
6		2	4	11		2	9

EXERCISES.

9 less 2?	9 less 1?	4 less 3?	5 less 2?
6 less 2?	4 less 1?	7 less 2?	5 and 4?
10 less 2?	3 less 2?	5 less 2?	8 and 6?
1 and 7?	4 less 2?	9 and 4?	6 less 2?
7 less 2!	3 less 1?	7 and 3?	8 less 2?
5 less 2?	10 less 2?	6 less 2?	9 less 2?

Lesson Fourteenth.—Decreasing by Threes.

1. Three balloons were sent up; if they all come down will any remain up?
2. There are four bottles of ink; should three of them be emptied, how many will still be full?
3. Five rabbits were in a cage; three got out; how many remained?
4. A little girl who had six cents spent three; how many were left?
5. Seven buggies are going by; if three of them stop how many will continue on?
6. Katie had eight peaches, but she gave three of them to her brother; how many did she keep?
7. Eddie, having nine chestnuts, gave three to George; how many did he have left?
8. When we eat three of the ten melons which father bought, how many will remain?
9. A merchant had eleven pitchers; after three were sold, how many remained?
10. Twelve spools of cotton are in a paper; how many will be left after three are used?

THE TABLE.

3	— 3 =	0		8	— 3 =	5
4	3	1		9	3	6
5	3	2		10	3	7
6	3	3		11	3	8
7	3	4		12	3	9

EXERCISES.

7 less 2?	13 less 3?	6 less 2?	5 and 4?
9 less 3?	5 less 2?	12 less 3?	4 less 1?
8 and 5?	6 less 2?	7 less 2?	11 less 3?
9 less 6?	8 less 3?	10 and 4?	7 and 3?
7 less 3?	4 less 3?	5 less 3?	8 and 6?
9 less 3?	5 less 2?	6 less 3?	9 less 2?

Lesson Fifteenth.—Decreasing by Fours.

1. Henry had four pet birds; having given them away he wanted to know how many were left?

2. A new house is to have five chimneys, but four of which are finished; how many are unfinished?

3. A tanner sells four hides out of six; how many remain?

4. Sarah goes to school four days in the week; how many does she stay at home?

5. Eight girls are sitting in the parlor; if four walk out, how many remain?

6. A drover has nine mules, of which he sells a team of four; how many does he keep?

7. Ten hogsheads of sugar are at the depot; a dray hauls off four; how many are left?

8. Eleven geese are swimming in the pond; if four come out, how many will remain?

9. Twelve shoes are in a box; if four are sold, how many will there be in the box?

10. If four grapes are taken from a bunch of thirteen, how many will remain?

THE TABLE.

4	— 4	= 0		9	— 4	= 5
5	4	1		10	4	6
6	4	2		11	4	7
7	4	3		12	4	8
8	4	4		13	4	9

EXERCISES.

9 less 5?	4 and 3?	6 less 3?	5 less 2?
7 less 5?	3 and 3?	5 less 4?	9 less 2?
6 less 3?	3 less 1?	10 less 4?	8 and 6?
9 less 4?	7 less 4?	11 less 3?	6 less 4?
7 and 6?	8 less 4?	13 less 4?	5 and 7?
4 and 9?	5 less 3?	10 less 8?	12 less 4?

Lesson Sixteenth.—Decreasing by Fives.

1. Five pitchforks were in the tool house, but Jim threw them all out; how many remained?
2. There are six padlocks, five of which have keys; how many are without keys?
3. A ladder has seven rounds, five of which are painted; how many are not painted?
4. Eight girls are out walking; five have pink scarfs; how many have not?
5. A large mouth ate five out of nine dumplings; how many were left?
6. Mary had ten strings of chinquepins; she gave five to Sarah; how many were left?
7. Eleven boys are playing town-ball; if five are on one side how many are on the other?
8. Twelve bags of flour; after five were sold, how many remained?
9. Thirteen hands are sent to work on the road; five of them lie down in the shade; how many work?
10. Five men out of fourteen are six feet high; how many are less?

THE TABLE.

$$5 - 5 = 0 \qquad 10 - 5 = 5$$
$$6 \quad 5 \quad 1 \qquad 11 \quad 5 \quad 6$$
$$7 \quad 5 \quad 2 \qquad 12 \quad 5 \quad 7$$
$$8 \quad 5 \quad 3 \qquad 13 \quad 5 \quad 8$$
$$9 \quad 5 \quad 4 \qquad 14 \quad 5 \quad 9$$

EXERCISES.

14 less 5?	10 less 5?	15 less 5?	6 less 3?
6 and 4?	13 less 5?	7 and 3?	9 less 4?
11 less 5?	9 and 8?	12 less 5?	7 less 5?
9 less 5?	7 less 6?	6 less 5?	4 less 3?
6 and 5?	8 less 5?	10 less 5?	7 less 5?
4 and 9?	10 less 4?	5 and 6?	2 and 8?

Lesson Seventeenth.— Decreasing by Sixes.

1. Six rats made their nests under a stack of fodder; the cat caught them all; how many were left?

2. If we work six days out of seven, on how many do we rest?

3. Eight pigs are in the pen; if six are spotted, how many are not spotted?

4. Of nine bee-hives six are filled with honey; how many are not filled?

5. Lizzie has painted ten pictures; six have frames; how many have not?

6. Birdy is playing with eleven keys; if he keeps six, how many will he lose?

7. Twelve soldiers are marching; if there are six in the front rank, how many will there be in the rear rank?

8. If six flowers are pulled from a boquet of thirteen, how many remain?

9. Fourteen cows are in the pea field; six are lying down; how many are standing?

10. Fifteen servants are dancing to the banjo; six are men; how many are women?

THE TABLE.

6	—	6	=	0		
7		6		1		
8		6		2		
9		6		3		
10		6		4		

11	—	6	=	5
12		6		6
13		6		7
14		6		8
15		6		9

EXERCISES.

9 less 6?	11 less 6?	7 and 6?	18 less 5?
14 less 6?	10 less 6?	5 and 6?	12 less 6?
7 less 4?	10 less 5?	8 less 3?	6 less 6?
7 less 6?	6 and 9?	14 less 6?	6 and 6?
1 and 8?	9 less 6?	12 less 5?	15 less 6?
6 and 5?	13 less 6?	7 and 5?	9 less 4?

Lesson Eighteenth.—Decreasing by Sevens.

1. Seven cakes were on a plate; John ate them up; how many remained?

2. Eight pine trees grow together; seven are boxed for turpentine; how many are not?

3. Nine servants are picking cotton; if seven are women, how many are men?

4. Ten stacks of fodder are in the field; seven are fenced around; how many are not?

5 Out of eleven large yams seven have been cut with the hoe; how many are uncut?

6. Seven out of twelve cabins are double; how many are single?

7. In a family of thirteen children seven are boys; how many are girls?

8. Fourteen loads of wood have been hauled, of which seven were light-wood knots; how many were not?

9. Fifteen martin gourds are upon a pole; seven are empty; how many are occupied?

10. Of sixteen bales of cotton, if seven are hauled at one load, how many would be left for another?

THE TABLE.

7	— 7	= 0	12	— 7	= 5
8	7	1	13	7	6
9	7	2	14	7	7
10	7	3	15	7	8
11	7	4	16	7	9

EXERCISES.

16 less 7?	15 less 6?	10 less 5?	4 and 9?
9 less 7?	17 less 7?	13 less 6?	6 and 8?
5 and 6?	8 less 7?	9 less 5?	12 less 7?
9 and 4?	7 and 7?	16 less 7?	•14 less 4?
8 and 4?	10 less 7?	12 less 7?	7 and 9?
6 and 7?	15 less 7?	11 less 7?	13 less 7?

Lesson Nineteenth.—Decreasing by Eights.

1. A boy who had eight jackets but had worn them all out; how many were left?

2. Nine bacon sides were hanging in the smoke house; eight have been given out; how many are left?

3. James bought ten plugs of tobacco; after chewing eight how many were left?

4. Bought eleven dozen eggs; eight dozen only were fresh; how many were spoiled?

5. Take eight bales of cotton from twelve bales, and how many will be left?

6. Thirteen girls were in two classes; if eight are in one how many are there in the other?

7. Fourteen spinning wheels are at the shop; when eight are sold how many will be left?

8. Sarah has fifteen books; eight of them contain pictures; how many do not?

9. Sixteen hogs are in the pen; if eight are killed how many will be left?

10. Seventeen dead trees were in the field, but the wind blew down eight; how many are standing?

THE TABLE.

8	—	8	=	0		13 — 8 = 5
9		8		1		14 8 6
10		8		2		15 8 7
11		8		3		16 8 8
12		8		4		17 8 9

EXERCISES.

16 less 8?	9 less 4?	13 less 8?	15 less 7?
15 less 8?	7 and 9?	14 less 7?	12 less 8?
10 less 8?	13 less 7?	6 and 9?	11 less 8?
6 and 8?	9 less 5?	13 less 7?	17 less 8?
15 less 8?	10 less 8?	12 less 7?	9 less 8?
12 less 6?	7 and 9?	10 less 3?	14 less 8?

Lesson Twentieth.—Decreasing by Nines.

1. Nine piles of brush are burning; when they are all out, how many will remain?

2. Nine out of ten jars of lard have been used; how many remained?

3. If a fence is eleven rails high and nine of them are old, how many are new?

4. Twelve chairs are in a room; if nine are split bottomed, how many are not?

5. At my quarters there are thirteen chimneys, nine of which are built with sticks; how many are not?

6. Fourteen shuck pens stand in a row; nine are full; how many have been emptied?

7. Out of fifteen days John rode to school nine times; how often did he walk?

8. Sixteen plows are at the shop; nine are scooters; how many are not?

9. Seventeen plates were in a tray; Sallie upset the tray and nine were broken; how many were not?

10. Eighteen sacks of salt were on the wagon; nine have been carried into the smoke house; how many are left?

THE TABLE.

9	— 9	= 0		14	— 9	= 5
10	9	1		15	9	6
11	9	2		16	9	7
12	9	3		17	9	8
13	9	4		18	9	9

EXERCISES.

15 less 9?	16 less 8?	8 less 2?	9 less 9?
5 and 4?	12 less 9?	9 and 5?	11 less 9?
17 less 8?.	16 less 9?	19 less 9?	15 less 6?
12 less 8?	18 less 9?	4 and 6?	10 less 9?
13 less 9?	9 and 5?	17 less 9?	7 and 6?

Lesson Twenty-first.—Decreasing by Tens.

1. Ten boys were playing ball; when they all stopped how many were still playing?

2. Eleven pigeons have alighted on their house; if ten fly off, how many will be left?

3. Twelve wagons are on the road; if ten of them have loads how many have not?

4. Thirteen shawls are hanging in a row; if ten of them are removed, how many will remain?

5. Take ten pods of pepper from a string of fourteen pods, how many will remain?

6. A girl who has fifteen dimes spends ten; how many are left?

7. John is sixteen years old; Henry is ten years; required the difference of their ages?

8. If ten pupils stay at home from a class of seventeen, how many will be present?

9. Eighteen fruit trees are in the garden; ten of them are in bloom; how many are not?

10. The distance between two places is nineteen miles; if we have traveled ten, how many more must we travel?

THE TABLE.

10	— 10	= 0	15	— 10	= 5
11	10	1	16	10	6
12	10	2	17	10	7
13	10	3	18	10	8
14	10	4	19	10	9

EXERCISES.

18 less 10?	15 less 9?	12 less 6?	17 less 10?
9 and 7?	15 less 8?	16 less 10?	18 less 9?
14 less 7?	17 less 9?	13 less 10?	10 less 7?
12 less 10?	19 less 6?	14 less 10?	10 less 4?
8 and 6?	13 less 9?	17 less 8?	19 less 10?
10 less 10?	17 less 9?	12 less 6?	15 less 8?

Lesson Twenty-second.—Miscellaneous Questions.

1. Henry had eight peaches, but he gave three to his sister; how many did he have left?

2. Mary had ten dimes in her purse, but she lost four of them; how many had she left?

3. The dog had nine puppies, but six of them were drowned; how many had she left?

4. The hen had fourteen chickens, but she raised only six of them; how many did she lose?

5. Katie had thirteen chestnuts, of which eight were roasted; how many were not roasted?

6. Eighteen girls went up to recite, but seven were sent to their seats for not knowing their lessons; how many remained?

7. Sarah has twelve apples; she gives two to George and two to Birdy; how many has she for herself?

8. Carrie had nine pencils; she lost three and gave one each to her two little brothers; how many are left?

9. Twelve eggs are boiling; four are taken out by James and three by John; how many are left?

10. Mary has three pies, but gives away two peaches; what has she left? Why?

11. John found six marbles and his mother gave him three more; but he owed fifteen marbles; how many did he still need?

12. There are fourteen sheets of paper on the table; two of them are yellow, five are brown, and the rest are white; how many are white?

EXERCISES.

$6+4+3-2$?	$12-3-2-4$?	$4+2+1+5+3$?
$5+6+4-2$?	$14-5-3+6$?	$5+3+2+3+4$?
$4+7+3-4$?	$15+3-9-2$?	$3+2+4+1+3$?
$5+2+4-3$?	$13+6-4+3$?	$2+4+3+2+1$?
$3+7-2+4$?	$12-5+4-2$?	$8+1+3+4+2$?
$8+5-4+3$?	$11+6-2-3$?	$3+5+2+5+4$?
$5+5-3+4$?	$14-3+9-2$?	$5+2+3+4+5$?
$6+9-2-4$?	$16-2-3+4$?	$3+4+2+5+6$?
$9+3-5+2$?	$5+4+6+3$?	$2+2+4+4+3$?
$8+6-3-4$?	$7+9+3+6$?	$5+4+3+6+2$?
$9+4-2-5$?	$4+9-2+5$?	$4+5+3+5+2$?
$8-3+4+6$?	$13-4+6-7$?	$6+3+2+4+1$?
$7-4+5+8$?	$12+6-4-2$?	$8+2+1+3+2$?
$9-6+4-3$?	$15+4-3-4$?	$9+3+4+5+3$?
$8-3+4+5$?	$8+9+4-5$?	$3+4+2+4+3$?
$7+5-3-4$?	$9+8+7-3$?	$6+3+2+4+5$?
$8+6-5+6$?	$4-9+2+4$?	$4+5+3+6+2$?

NOTE.—Let not the Teacher be alarmed at this array of figures, or lay it aside without faithful trial. The author, in this little Primer, merely indicates the outline of his course. A great many additional examples should be made by the Teacher at every step in the progress of his class.

MULTIPLICATION.

To the Teacher.

Multiplication consists in successive additions of the same number. Availing himself of this fact, the author has adopted, with great success, the following plan for teaching the multiplication table to young children :

The floor of the room is supposed to be divided into nine compartments, thus:

A tenth compartment is supposed to be on the ceiling directly over the middle one on the floor.

In each of these compartments a digit is to be written with chalk. The cypher being supposed to be written in the compartment on the ceiling, thus :

1	2	3
4	5	6
7	8	9

Every thing is now ready. Begin by counting. The class may be placed on the line indicated by the figures 2, 5, 8, with their faces toward the line 3, 6, 9. Pointing to each figure as they advance, the class repeats, in concert, as follows : 3, 6, 9, 12, 15, 18, 21, 24, 27, 30. Care

2

must be taken to make the pupils understand that, when adding by threes, they cannot go from 9 to 2, that the 2 is only the final figure and the number becomes 12; so, from 18 to 1, the 1 becomes 21, and so on.

It is neither desirable nor necessary to stop the process at thirty. Continue thus: 33, 36, 39, 42, &c., at the teacher's pleasure. *Always count backwards* from whatever number you may have ascended to.

What has been stated above as to counting by threes, holds good for the other odd numbers, except five. Arrange the class always to front towards the line in which the number they are to increase by is found. The number five is so simple a one to count by that no directions are necessary for it.

To count by the even numbers, let the class face the number, and five motions of the arm will always point out the proper unit figure in the number required. These five motions are: first, to the front; second, to the left; third, to the right; fourth, to the rear; and fifth, to the 0.

Illustration: Suppose it is required to count by twos. The diagram on the floor may then be represented thus:

1	2	3
4	5	6
7	8	9

The exercise will proceed: front, 2; left, 4; right, 6; rear, 8; overhead, 10; front, 12; left, 14; right, 16; rear 18; overhead 20, and so on. Practice counting forwards and backwards.

It may be required to begin with an odd number and increase by an even number. Five motions of the arm are again sufficient, viz: Two in front, one underfoot, and two in the rear, (first, front left; second, front right; third, downwards; fourth, rear left; fifth, rear right.) These movements describe the letter Z.

These exercises should proceed daily until each pupil can readily count forwards and backwards to the number of twenty additions without the motion of the arm. These motions and the numbers on the floor being mere scaffolding, should be dispensed with as soon as possible.

But what has all this to do with multiplication? Everything with multiplication and everything with division. Instead of counting 4, 8, 12, 16, 20, and so on, let the class begin in concert to add the multiplication formula,* which is all that is required; thus: Once four is four; two times four are 8; three times four are 12, and so on.

Instead of counting backwards thus: 30, 24, 18, 12, 6, add the division formula thus: 6 into 30, five times; 6 into 24, four times; 6 into 18, three times; 6 into 12 two times; 6 into 6, once.

The slate and the blackboard should be brought into frequent requisition in all this training. If the pupil frequently recites out the columns (2, 3, 4, &c.,) he will become much more perfectly acquainted with them than by any other process.

---

*The formula is elliptical; expressed fully it becomes four taken five times becomes (are) twenty. That is, abbreviated, five times four are twenty.

MULTIPLICATION.

Lesson Twenty-third.—Increasing by Twos.

1	2	3
4	5	6
7	8	9

1. At one cent each what will be the cost of two apples?
2. At two cents each required the cost of two peaches.
3. If one cap costs two dollars, how much will three caps cost?
4. If one gun costs two dollars, what will four guns cost?
5. Required the cost of five pencils at two cents each.
6. Required the cost of six books at two dollars each.
7. How much will seven buckles cost at two cents each?
8. One pair of shoes cost two dollars; required the cost of eight pairs.
9. One hammer costs two dimes; what will nine hammers cost?
10. Ten tops will cost how much, if one cost two cents?

THE TABLE.

1	× 2 =	2		6	× 2 =	12
2	2	4		7	2	14
3	2	6		8	2	16
4	2	8		9	2	18
5	2	10		10	2	20

EXERCISES.

2 times 3?	4 times 2?	7 times 6?	2 times 9?
4 times 2?	8 times 1?	2 times 7?	8 times 2?
2 times 7?	9 times 2?	5 times 2?	3 times 2?
3 times 2?	6 times 2?	4 times 2?	6 and 9?
10 times 2?	8 times 2?	2 times 7?	12 less 9?
13 less 6?	2 times 6?	4 and 8?	3 times 2?
4 times 2?	8 times 2?	9 and 7?	7 times 2?

Lesson Twenty-fourth.——Increasing by Threes.

3	6	9
2	5	8
1	4	7

1. What cost three marbles at one cent each?

2. How much will two pens cost at three cents each?

3. If three boys each recite three lessons, how many will they all recite?

4. Four girls have each three dolls; how many have they all?

5. Three bottles of ink at five cents each will cost how much?

6. There are three nests, each containing six eggs; how many eggs in all three?

7. If there are seven pounds of butter in each of three boxes, how many pounds will that make?

8. Eight boys have each three cents; how many have they all?

9. Three oranges at nine cents apiece will cost how much?

10. In one dime there are ten cents; how many cents in three dimes?

THE TABLE.

1	× 3	= 3		6	× 3	= 18
2	3	6		7	3	21
3	3	9		8	3	24
4	3	12		9	3	27
5	3	15		10	3	30

EXERCISES.

3 times 3?	5 and 6?	1 times 3?	4 times 3?
9 times 2?	2 times 3?	8 times 3?	1 and 7?
7 times 3?	8 times 3?	12 less 9?	6 times 3?
16 and 9?	4 times 2?	5 times 3?	9 times 3?
16 less 8?	3 times 9?	4 times 3?	7 times 3?
1 and 7?	9 times 3?	11 less 5?	8 less 3?
3 times 2?	9 times 4?	5 times 3?	5 times 8?

Lesson Twenty-fifth.---Increasing by Fours.

7	4	1
8	5	2
9	6	3

1. Four tumblers each holding one pint would contain how much?

2. Two rooms each contain four windows, have how many in all?

3. Three cats have each four legs; how many legs have they all?

4. Four vests are each to have four buttons; how many will be required?

5. How many pencils will there be in five boxes if each box contains four?

6. How many horses are there in four teams if each team consists of six horses?

7. Seven trees are in a row, and there are four rows; how many trees will that make?

8. What is the joint age of four girls who are each eight years old?

9. A four sided pig-pen has nine rails in each side; how many rails?

10. In a dollar there are ten dimes; how many dimes are there in four dollars?

THE TABLE.

1	× 4 =	4		6	× 4 =	24
2	4	8		7	4	28
3	4	12		8	4	32
4	4	16		9	4	36
5	4	20		10	4	40

EXERCISES.

7 times 4?	6 times 3?	3 times 4?	4 times 6?
11 less 9?	4 times 4?	1 times 4?	7 times 4?
5 times 4?	9 times 3?	8 times 4?	9 and 7?
9 times 4?	4 times 4?	5 and 9?	9 times 6?
2 times 4?	5 times 4?	6 times 4?	8 times 5?

Lesson Twenty-Sixth---Increasing by Fives.

5

1. At one dollar each what will five capes cost?

2. At five dollars each two bonnets will cost how much?

3. If cotton thread costs five cents a spool, how much will three spools cost?

4. If four plates contain each five potatoes, how many do they all contain?

5. What cost five oranges at five cents each?

6. Required the cost of five sheep at six dollars apiece?

7. How much will seven guns cost at five dollars each?

8. Every boy ought to have eight fingers; how many ought five boys to have?

9. Five clusters each contain nine pods; how many pods will the five contain?

10. Five piles each contain ten dollars; how many dollars in all?

THE TABLE.

1	×	5	=	5	6 × 5 =	30
2		5		10	7 5	35
3		5		15	8 5	40
4		5		20	9 5	45
5		5		25	10 5	50

EXERCISES.

2 times 2?	4 times 5?	9 times 5?	7 times 5?
8 times 4?	8 times 3?	7 times 4?	1 times 5?
5 times 5?	2 times 5?	3 times 5?	5 times 4?
4 times 4?	8 times 5?	6 times 5?	9 times 4?
6 times 5?	7 times 5?	4 times 5?	9 times 5?
4 and 9?	6 times 4?	3 times 4?	7 times 5?
4 times 5?	9 times 5?	13 less 2?	15 less 6?
5 times 5?	8 times 6?	10 times 3?	2 times 6?
3 times 4?	9 times 9?	8 times 9?	3 times 2?
8 times 3?	7 times 4?	5 times 4?	8 times 4?
3 and 9?	4 times 3?	16 less 9?	9 times 4?

Lesson Twenty-Seventh.---Increasing by Sixes.

3	6	9
2	5	8
1	4	7

1. Required the cost of six screws at one cent apiece.
2. At two dollars each what will six handkerchiefs cost?
3. If six boys can sit on a bench, how many can sit on three benches?
4. How much must I pay for four pounds of soap at six cents per pound?
5. One spool costs five cents; required the cost of six spools.
6. Six balls will cost how much at six cents apiece?
7. What number of yards in six dresses if there are seven yards in each?
8. In each wagon are six barrels; how many are there in eight wagons?
9. John gave nine marbles to each of six little boys; how many did he give away?
10. Ten mills make a cent; how many mills in six cents?

/ THE TABLE.

1	× 6 =	6		6	× 6 =	36
2	6	12		7	6	42
3	6	18		8	6	48
4	6	24		9	6	54
5	6	30		10	6	60

EXERCISES.

5 times 6 ?	4 times 6 ?	5 times 6 ?	9 and 5 ?
9 times 5 ?	9 times 2 ?	8 times 4 ?	7 times 6 ?
1 time 6 ?	3 times 5 ?	11 less 8 ?	6 times 6 ?
7 times 4 ?	6 times 6 ?	3 times 6 ?	4 times 5 ?
5 times 6 ?	4 times 4 ?	6 times 6 ?	2 times 6 ?

Lesson Twenty-Eighth.----Increasing by Sevens.

7	4	1
8	5	2
9	6	3

1. What will seven sticks of candy cost at one cent a stick?

2. How much will two suits of clothes cost at seven dollars a suit?

3. Three little girls had each seven chinquepins; how many did all have?

4. At the carriage shop are seven buggies, each with four wheels; how many wheels have they all?

5. There are seven days in one week; how many days in five weeks?

6. The merchant has seven boxes of ink, each containing six bottles; how many bottles has he?

7. Seven buttons on seven coats make how many?

8. If eight boys have each seven books, how many have they all?

9. How much will seven barrels of flour cost at nine dollars a barrel?

10. Ten dollars makes one eagle; how many dollars in seven eagles?.

THE TABLE.

1	×	7 =	7	6	× 7 =	42
2		7	14	7	7	49
3		7	21	8	7	56
4		7	28	9	7	63
5		7	35	10	7	70

EXERCISES.

6 times 7 ?	7 times 7 ?	5 times 7 ?	9 times 7 ?
3 times 6 ?	1 times 7 ?	3 times 7 ?	4 times 6 ?
9 times 6 ?	5 times 7 ?	14 less 5 ?	6 times 7 ?
4 times 7 ?	9 times 5 ?	8 times 7 ?	7 times 7 ?
4 times 6 ?	7 times 8 ?	2 times 7 ?	4 times 7 ?
9 times 9 ?	8 times 7 ?	5 times 6 ?	3 times 7 ?
5 times 7 ?	2 times 7 ?	7 times 7 ?	4 times 7 ?

Lesson Twenty-Ninth.----Increasing by Eights.

9	8	7
6	5	4
3	2	1

1. Eight cakes cost how much at one cent apiece ?

2. Two legs to each pair of tongs ; how many legs to eight pairs ?

3. What will eight penknives cost at three dollars apiece ?

4. If one cedar pencil costs eight cents, what will four cost ?

5. Required the cost of eight calves at five dollars each.

6. How much must I pay for six pumpkins at eight cents apiece?

7. There are seven rows of trees and eight in a row; how many trees?

8. On a checker-board there are eight rows of squares and eight in a row; how many squares?

9. A train of nine cars, and each car has eight wheels; required the whole number of wheels?

10. What will be the cost of eight brass clocks at ten dollars each?

THE TABLE.

1	×	8 = 8		6	×	8 = 48
2	8	16		7	8	56
3	8	24		8	8	64
4	8	32		9	8	72
5	8	40		10	8	80

EXERCISES.

7 times 8?	9 times 8?	3 times 8?	4 times 8?
9 times 8?	4 times 8?	1 times 8?	6 times 8?
5 times 7?	6 times 7?	8 times 7?	9 times 7?
4 times 7?	4 times 8?	5 times 8?	9 times 8?
2 times 8?	6 times 8?	9 times 8?	5 times 7?
8 times 8?	8 times 7?	3 times 8?	2 times 7?

Lesson Thirtieth.----Increasing by Nines.

9	8	7
6	5	4
3	2	1

1. Nine girls have each one cent; how many have they all?

2. If one urn costs nine dollars, how much will two urns cost?

3. One bottle of medicine costs three dollars; how much will nine bottles cost?

4. Four buckets contain each nine quarts of milk; how much do all the buckets contain?

5. Nine long words each of five syllables, contain how many syllables?

6. How much will six papers of needles cost at nine cents a paper?

7. Seven yards of trimming at nine cents per yard will cost how much?

8. What must I pay for nine cords of wood at eight dollars a cord?

9. Nine girls each learn nine hymns; how many did all learn?

10. Required the weight of ten lumps of butter, each weighing nine pounds.

THE TABLE.

1	+ 9 =	9		6	+ 9 =	54
2	9	18		7	9	63
3	9	27		8	9	72
4	9	36		9	9	81
5	9	45		10	9	90

EXERCISES.

7 times 9?	5 times 8?	3 times 9?	8 times 8?
3 times 8?	8 times 9?	5 times 7?	4 times 9?
6 times 7?	9 times 7?	6 times 8?	9 times 9?
4 times 9?	2 times 9?	4 times 9?	7 times 7?
5 times 9?	6 times 9?	9 times 7?	8 times 7?
3 times 7?	4 times 8?	6 times 7?	9 times 8?
7 times 9?	3 times 8?	5 times 9?	8 times 7?

Lesson Thirty-first.----Miscellaneous Questions.

1. Three little girls have each four fingers on each hand ; how many fingers have they all?

2. How many legs have seven horses?

3. I have three rakes with eight teeth each, how many teeth have they all?

4. How many hands and feet have six little girls?

5. There are four windows in my room, each with two sash and six panes of glass in each sash. How many panes of glass are there in the room?

6. Eight flies have how many feet?

7. How many horns and feet ought eight cows to have?

8. A railroad car has four wheels on each side ; how many wheels have eight cars?

9. In my house there are two pictures on each wall of each room ; how many pictures are there in four rooms?

10. Two meeting-houses are lighted by four chan dol - iers, each supporting three lamps ; how many lamps in both?

11. I have five jackets each with seven buttons on a side; how many buttons on all the jackets?

12. Mary bought three boxes each with four cards containing six hooks and eyes? how many in all?

EXERCISES.

$4 \times 9 + 2$? *	$2 \times 2 \times 3$ †	$6 \times 5 - 2$
$8 \times 7 + 3$?	$2 \times 2 \times 4$	$5 \times 4 - 3$
$9 \times 5 + 1$?	$2 \times 2 \times 5$	$7 \times 9 - 4$
$8 \times 4 + 5$?	$3 \times 2 \times 4$	$8 \times 7 - 2$
$6 \times 2 + 8$?	$2 \times 3 \times 4$	$6 \times 8 - 3$
$8 \times 2 + 3$?	$4 \times 2 \times 3$	$9 \times 5 - 6$
$4 \times 6 + 5$?	$6 \times 2 \times 4$	$8 \times 9 - 3$
$8 \times 3 + 4$?	$9 \times 2 \times 3$	$9 \times 9 - 2$

* Either 4 times 9 and 2 to carry, or 4 times 9 and 2, or 4 times 9 and two ninths of nine.

† Read two times 2 times 3, (the last times pronounced tims.)

DIVISION.

1	2	3
4	5	6
7	8	9

1. How often can two apples be taken from two apples?

2. How often will two go into two?

3. Four apples are to be divided between two boys, how many will each one get?

4. Six cents will buy how many pencils at two cents apiece?

5. How many cakes at two cents apiece can I buy for eight cents?

6. Ten dollars to be divided between two persons; how much will each receive?

7. Twelve girls sit on two benches; how many are there on each?

8. Divide fourteen books between two boys; how many will each receive?

9. Sixteen finger rings are to be divided between two girls; how many will each receive?

10. Eighteen lights of glass in two sashes; how many in each?

11. Twenty dollars in two equal notes; how many in each note?

THE TABLE.

2 ÷ 2 = 1			12 ÷ 2 = 6		
4	2	2	14	2	7
6	2	3	16	2	8
8	2	4	18	2	9
10	2	5	20	2	10

EXERCISES.

2 into 6?	2 into 2?	3 times 9?	2 into 6?
2 into 4?	2 into 10?	2 into 12?	2 into 8?
5 times 4?	5 times 7?	5 times 8?	2 into 14?
9 times 5?	2 into 18?	3 times 9?	2 into 16?
7 times 9?	5 times 8?	9 times 7?	2 into 14?
5 times 6?	2 into 14?	9 times 8?	5 into 15?

Lesson Thirty-third.—Increasing by Threes.

3	6	9
2	5	8
1	4	7

1. How often will three go into three?

2. If you divide six hair-pins between three girls, how many will each one receive?

3. A father wants to distribute nine books among his three children; how many will each child receive?

4. Twelve cherries will make how many bunches of three each?

5. Three squirrels eat fifteen nuts; how many does each one eat?

6. Eighteen boys are to be seated on three benches; how many should sit on each bench?

7. Twenty-one days make three weeks; how many days in one week?

8. Twenty-four wheels are to three cars; how many wheels is that to each?

9. Three men shoot each the same number of birds; together they have twenty-seven. How many did each man shoot?

10. Thirty dimes are in three piles; how many are there in each?

THE TABLE.

3	÷	3	= 1	18 ÷ 3	=	6
6		3	2	21	3	7
9		3	3	24	3	8
12		3	4	27	3	9
15		3	5	30	3	10

EXERCISES.

3 into 9?	3 into 15?	3 into 27?	4 times 5?
3 into 6?	3 into 3?	3 into 24?	9 times 7?
3 into 9?	3 into 18?	3 into 30?	3 times 9?
2 into 8?	3 into 30?	2 into 18?	5 times 9?
2 into 4?	2 into 18?	2 into 12?	5 times 8?

Lesson Thirty-fourth.—Increasing by Fours.

7	4	1
8	5	2
9	6	3

3

1. Four boys divide four balls between themselves; how many is that to each?

2. Eight wheels are sufficient for how many carriages?

3 Divide twelve marbles among four boys; required each one's share?

4. Sixteen legs are sufficient to make how many tables, each with four legs?

5. Divide twenty dollars equally between four poor families; what amount will each receive?

6. Four wagons haul twenty-four bales of cotton; how many bales to each wagon?

7. Twenty-eight soldiers are marching in ranks of four each; how many soldiers in each rank?

8. Thirty-two pounds of sugar will fill how many four pound bags?

9. Four houses have thirty-six rooms; how many rooms in each house?

10. Forty quarters are equal to how many dollars?

THE TABLE.

4	÷	4	=	1		
8		4		2		
12		4		3		
16		4		4		
20		4		5		

24 ÷ 4 = 6
28   4   7
32   4   8
36   4   9
40   4   10

EXERCISES.

4 into 4?	7 into 28?	4 into 12?	2 into 16?
3 into 3?	3 into 24?	4 into 16?	4 into 12?
4 into 8?	3 into 15?	4 into 24?	2 into 18?
4 into 20?	3 into 18?	4 into 36?	4 into 32?
3 into 27?	3 into 15?	4 into 40?	4 into 36?
4 into 32?	2 into 20?	6 into 18?	5 into 50?
4 into 28?	5 into 30?	9 into 81?	7 into 56?

Lesson Thirty-fifth.—Increasing by Fives.

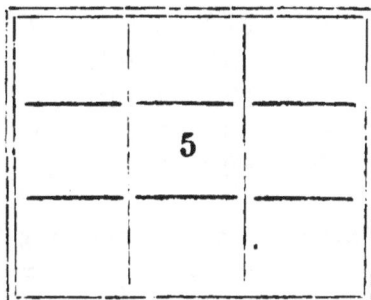

1. Five boys have five canes; how many have each ?
2. Five cows have ten horns; how many have each?
3. Divide fifteen pencils among five scholars; how many will each get ?
4. Twenty dollars will buy how many hats at five dollars each?
5. Twenty-five cents are exchanged for five equal change bills; required the value of each ?
6. Distribute thirty cords of wood equally between five poor families; how many cords to each ?
7. At five cents a skein, how many skeins of silk will thirty-five cents buy ?
8. Forty chairs are to be carried in one trip by five men; how many chairs must each one carry ?
9. Five drays are to haul forty-five barrels in one trip; how many to each dray?
10. Exchange a fifty dollar note for five equal smaller notes; required the value of each?

THE TABLE.

5	÷ 5	= 1		30	÷ 5	= 6
10	5	2		35	5	7
15	5	3		40	5	8
20	5	4		45	5	9
25	5	5		50	5	10

EXERCISES.

5 into 15?	3 into 27?	5 into 30?	5 into 5?
4 into 24?	5 into 25?	5 into 10?	5 into 40?
5 into 35?	5 into 15?	4 into 12?	3 into 24?
5 into 45?	5 into 20?	3 into 18?	5 into 25?
4 into 36?	2 into 10?	3 into 24?	3 into 30?

Lesson Thirty-sixth.—Increasing by Sixes.

3	6	9
2	5	8
1	4	7

1. How often will six go into six?
2. Six boys have twelve thumbs; how many have each?
3. Eighteen garments are to be divided between six men; how many to each?
4. Six silver forks have twenty-four tines; how many has each fork?
5. Thirty dollars pay for six pairs of boots; required the price of one pair?
6. Thirty-six chairs are in six rows; how many are in each row?
7. The joint ages of six girls of equal age was forty-two years; what was the age of each?
8. Forty-eight lots of land are to be divided equally between six heirs; how many does each one receive?
9. Fifty-four legs are sufficient to furnish how many flies?
10. Sixty acres of land are to be laid off in six lots; how many acres will each contain?

THE TABLE.

6	÷ 6	= 1		36	÷ 6	=	6
12	6	2		42	6		7
18	6	3		48	6		8
24	6	4		54	6		9
30	6	5		60	6		10

EXERCISES.

6 into 24 ?	2 into 16 ?	6 into 12 ?	3 into 21 ?
6 into 30 ?	6 into 18 ?	6 into 6 ?	6 into 24 ?
5 into 35 ?	3 into 18 ?	6 into 42 ?	5 into 45 ?
4 into 24 ?	3 into 9 ?	6 into 48 ?	3 into 27 ?
4 into 16 ?	5 into 50 ?	6 into 54 ?	6 into 48 ?
4 into 12 ?	3 into 21 ?	5 into 45 ?	6 into 42 ?

Lesson Thirty-seventh.---Increasing by Sevens.

·7	4	1
8	5	2
9	6	3

1. How many sevens are contained in seven ?

2. A dress may be made of seven yards of homespun ; how many dresses will fourteen yards make?

3. Twenty-one days make how many weeks?

4. Seven men divide twenty-eight mules equally among themselves ; how many does each man receive?

5. Thirty-five acres are to be plowed by seven men ; what should be the task of each, supposing each to plow the same number of acres ?

6. If seven pupils sit on one bench, how many benches will forty-two pupils need?

7 Forty-nine dinners should be eaten in how many weeks?

8. In seven hymns of equal length are fifty-six verses; how many verses have each?

9 In digging over seven hills, I found sixty-three large yams; how many was that to the hill?

10. Seven boys shoot seventy squirrels; how many was that apiece?

THE TABLE.

$7 \div 7 = 1$			$42 \div 7 = 6$		
14	7	2	49	7	7
21	7	3	56	7	8
28	7	4	63	7	9
35	7	5	70	7	10

EXERCISES.

7 into 21?	7 into 42?	7 into 63?	6 into 24?
7 into 35?	7 into 56?	7 into 7?	7 into 28?
7 into 49?	6 into 30?	7 into 21?	7 into 42?
7 into 63?	7 into 14?	7 into 35?	6 into 36?
3 into 21?	7 into 28?	7 into 49?	7 into 70?
6 into 48?	4 into 36?	3 into 27?	5 into 45?

Lesson Thirty-eighth.---Increasing by Eights.

9	8	7
6	5	4
3	2	1

1. How many times does eight contain eight?
2. There are sixteen blades in eight knives; how many has each?
3. Twenty-four tierces of rice will make how many loads, eight tierces being a load?
4. Suppose a rake has eight teeth; how many rakes would thirty-two teeth furnish?
5. A flute has eight keys; how many such flutes would forty keys furnish?
6. Sixty-four books are to be placed upon shelves that hold eight books each; how many shelves will be filled?
7. If there are eight horses to a team; how many teams would fifty-six horses make?
8. On a draughts board there are sixty-four squares, and eight squares in a row; how many rows are there?
9. Seventy-two wheels will suffice for how many cars; eight wheels being to one car?
10. Eighty dollars are in eight gold pieces; required the value of each?

THE TABLE.

8	÷	8	=	1	48 ÷ 8 = 6
16		8		2	56 8 7
24		8		3	64 8 8
32		8		4	72 8 9
40		8		5	80 8 10

EXERCISES.

8 into 32?	6 into 36?	8 into 24?	8 into 8?
7 into 63?	6 into 24?	8 into 40?	8 into 72?
8 into 64?	8 into 48?	8 into 16?	8 into 80?
7 into 35?	3 into 27?	8 into 64?	8 into 48?
8 into 56?	8 into 32?	7 into 49?	6 into 48?
3 into 27?	5 into 45?	6 into 30?	9 into 72?
4 into 28?	8 into 64?	7 into 49?	6 into 60?

Lesson Thirty-Ninth---Increasing by Nines.

9	8	7
6	5	4
3	2	1

1. How many times is nine contained in nine?

2. Place eighteen chairs, nine in a row; how many rows will be made?

3. Nine families have twenty-seven children; how many is that for each?

4. Thirty-six spools of cotton are to be divided between nine women; how many are given to each?

5. Forty-five kegs of powder are to be distributed to nine regiments; how many kegs is that to each?

6. Fifty-four bales of cotton are hauled to market by nine wagons; how many bales to the wagon?

7. If board is nine dollars a week; how many weeks will sixty-three dollars pay for?

8. Nine merchants buy seventy-two hogsheads of sugar; how many apiece?

9. Eighty-one trees stand in nine rows; how many trees are there in each row?

10. How many cents in one dime, if nine dimes contain ninety cents?

THE TABLE.

9	÷ 9 =	1		54	÷ 9 =	6
18	9	2		63	9	7
27	9	3		72	9	8
36	9	4		81	9	9
45	9	5		90	9	10

9 into 18 ?	3 into 27 ?	6 into 36 ?	9 into 45 ?
8 into 64 ?	9 into 63 ?	8 into 32 ?	9 into 81 ?
9 into 90 ?	9 into 9 ?	7 into 21 ?	9 into 72 ?
7 into 63 ?	9 into 54 ?	7 into 35 ?	8 into 56 ?
9 into 27 ?	8 into 48 ?	7 into 49 ?	8 into 32 ?
9 into 81 ?	9 into 63 ?	6 into 24 ?	9 into 45 ?

DIVISION.

Lesson Fortieth.----Miscellaneous Questions.

1. A father bought sixty-three marbles to divide among his seven boys; how many should each one receive ?

2. Nine girls agree to knit seventy-two pairs of gloves for poor children ; how many must each one knit ?

3. Henry brought home eighteen chestnuts; how many burrs must he have cracked to get them, if he found two chestnuts in each burr ?

4. A cruel boy had forty-two fly legs; how many flies must he have killed to get them ?

5. One dray can haul seven barrels of flour ; how many such drays will be required to haul fifty-six barrels at one trip ?

6. I see sixty-four fingers ; how many girls must they belong to ? How many hands ?

7. My wagons require six mules to the team ; thirty mules will be sufficient for how many wagons ?

8. At nine lights of glass to a sash, how many sash will seventy-two lights fill ?

9. Forty-five toes belong to how many feet ?

10. Eighty-one soldiers will make how many ranks of nine each ?

11. If there are seven yards in a dress, how many dresses will a piece of cloth measuring forty-two yards make ?

12. Sixty four fruit trees are to be planted in an orchard; how many rows will they make if there are eight trees in a row?

13. How often will four go into 12? 36? 28? 24? 8? 16?

14. 5 into 25? 40? 35? 20? 15? 35? 45?

15. 6 into 42? 30? 18? 54? 36? 48? 24? 12?

16. 3 into 15? 6? 21? 27? 9? 18? 12? 24?

$$42 - 2 \div 4? \; * \qquad 13 + 3 \div 2 \times 2? \; \dagger$$
$$57 - 3 \div 6? \qquad 9 \times 4 \div 3 \times 2?$$
$$45 + 4 \div 7? \qquad 8 \times 5 \div 2 \times 2?$$
$$32 + 4 \div 9? \qquad 2 \times 9 \div 3 \times 3?$$
$$54 + 2 \div 8? \qquad 5 \times 4 \div 2 \times 5?$$
$$75 - 3 \div 9? \qquad 6 \times 4 \div 4 \times 2?$$
$$87 - 6 \div 9? \qquad 6 \times 6 \div 3 \times 3?$$
$$48 + 6 + 6? \qquad 8 \times 6 \div 2 \times 2?$$
$$70 - 6 \div 8? \qquad 23 + 5 \div 2 \times 2?$$

---

* May be read two from forty-two leaves how many times four.
† May be read, divide two times two into thirteen and three.

MISCELLANEOUS TABLES.

TABLE OF UNITED STATES CURRENCY.

10 Mills	mako	1 Cent, '	marked	c.
10 Cents	"	1 Dime,	"	d.
10 Dimes	"	1 Dollar,	"	$.
10 Dollars	"	1 Eagle,	"	E.

TABLE OF ENGLISH MONEY.

4 Farthings	make	1 Penny,	marked	d.
12 Pence	"	1 Shilling,	"	s.
20 Shillings	"	1 Pound,	"	£.
21 Shillings sterling	"	1 Guinea,	"	G.
28 Shillings N. E.	"	1 Guinea,	"	G.

NOTE.—One pound Sterling is equal to $4.44 4-9 exchange value.

TABLE OF TROY WEIGHT.

24 Grains	make	1 Pennyweight,	marked	pwt.
20 Pennyweights	"	1 Ounce,	"	oz.
12 Ounces	"	1 Pound,	"	lb.

TABLE OF APOTHECARIES' WEIGHT.

20 Grains	make	1 Scruple.
3 Scruples	"	1 Dram.
8 Drams	"	1 Ounce.
12 Ounces	"	1 Pound.

Apothecaries mix their medicines by this weight; but buy and sell by Avoirdupois. The pound and ounce of this weight are the same as in Troy Weight.

TABLE OF AVOIRDUPOIS WEIGHT.

16 Drams	make	1 Ounce,	marked	oz.
16 Ounces	"	1 Pound,	"	lb.
28 Pounds	"	1 Quarter,	"	qr.
4 Quarters	"	1 Hundred Weight,	"	cwt.
20 Hundred Weight	"	1 Ton,	"	ton.

By this weight are weighed almost every kind of goods, and all metals except gold and silver. By a late law of Massachusetts, the cwt. contains 100 lbs. instead of 112 lbs.

A ton is reckoned at the custom-houses of the United States at 2240 lbs.

TABLE OF CLOTH MEASURE.

2¼	Inches	make	1 Nail,	marked		na.
4	Nails	"	1 Quarter of a yard,	"		qr.
4	Quarters	"	1 Yard,	"		yd.
3	Quarters	"	1 Ell Flemish,	"	E. F.	
5	Quarters	"	1 Ell English,	"	E. E.	
4	Quarters 1 1-3 inch	"	1 Ell Scotch,	"	E. S.	
6	Quarters	"	1 Ell French,	"	E. Fr.	

TABLE OF LONG MEASURE.

3	Barley-corns (*bc.*)	make	1 Inch,	marked		in.
12	Inches	"	1 Foot,	"		ft.
3	Feet	"	1 Yard,	"		yd.
5½	Yards, or 16½ Feet,	"	1 Rod or Pole,	"	rd. or po.	
40	Rods, or 220 Yds.	"	1 Furlong,	"		fur.
8	Furlongs	"	1 Mile,	"		mi.
3	Miles	"	1 League,	"		le.
60	Geographic, or ⎱ 69½ Statute Miles ⎰	"	1 Degree,	"	deg. or °	
360	Degrees,		the circumference of the earth.			

TABLE OF LAND OR SQUARE MEASURE.

144	Square In. (*sq. in.*)	make	1 Square Foot,	marked	sq. ft.
9	Square feet	"	1 Square Yard,	"	sq. yd.
30¼	Sq. Yds., or 272¼ ft.,	"	1 Sq. Rod or Pole,	"	sq. rod.
40	Sq. Rods or Poles	"	1 Rood,	"	R.
4	Roods	"	1 Acre,	"	A.
640	Acres	"	1 Square Mile,	"	sq. M.
16	Square Poles	"	1 Square Chain,	"	ch.
10	Square Chains	"	1 Acre,	"	A.

TABLE OF MEASURING DISTANCES.

7	92-100 Inches	make	1 Link.
25	Links	"	1 Pole.
100	Links	"	1 Chain.
10	Chains	"	1 Furlong.
8	Furlongs	"	1 Mile.

TABLE OF SOLID OR CUBIC MEASURE.

1728 Solid In. (*sol. in.*) make 1 Solid Foot,	marked sol. ft.
27 Solid Feet	" 1 Solid Yard,	" sol. yd.
40 ft. round tim., or } 50 ft. hewn timber	" 1 Ton,	" T.
16 Solid feet	" 1 Cord foot,	" c. ft.
8 Cord feet, or } 128 Solid feet	" 1 Cord of wood,	" cd.

A pile of wood 8 feet long, 4 feet wide, and 4 feet high, is a cord.

TABLE OF WINE MEASURE.

4	Gills (*gi*)	make 1 Pint,	marked pt.
2	Pints	" 1 Quart,	" qt.
4	Quarts	" 1 Gallon,	" gal.
31½	Gallons	" 1 Barrel,	" bbl.
42	Gallons	" 1 Tierce,	" tcc.
63	Gallons, or 2 barrels	" 1 Hogshead,	" hhd.
2	Tierces	" 1 Puncheon,	" pun.
2	Hogsheads, (126 gals.)	" 1 Pipe,	" pi.
2	Pipes, 4 hhds., or 252 gals.	" 1 Tun,	" T.

NOTE.—The Wine Gallon contains 231 cubic inches. Water-wine, and spirits, are measured and sold by this measure.

A cubic foot of distilled water weighs 1,000 ounces Avoirdupois.

The English Imperial Gallon contains 277¼ cubic inches, and weighs 10 lbs. Avoirdupois, or 12 lbs. 1 oz. 16 dwt. 16 gr. Troy. There is no legal measure in the United States for tierce, hogshead, puncheon, pipe, or butt.

TABLE OF DRY MEASURE.

2 Pints	make 1 Quart,	marked qt.
4 Quarts	" 1 Gallon,	" gal.
2 Gallons	" 1 Peck,	" pk.
4 Pecks	" 1 Bushel,	" bu.
36 Bushels	" 1 Chaldron,	" ch.

NOTE.—This measure is applied to all goods that are not liquid and are sold by measure, as corn, fruit, salt, coals, etc. A Winchester Bushel is 18½ inches in diameter, and 8 inches deep. The standard Gallon, Dry Measure, contains 268 4-5 cubic inches.

TABLE OF ALE AND BEER MEASURE.

2 Pints	make	1 Quart,	marked	qt.
4 Quarts	"	1 Gallon,	"	gal.
32 Gallons	"	1 Barrel,	"	bbl.
54 Gallous	"	1 Hogshead,	"	hhd.
2 Hogsheads	"	1 Butt,	"	butt.
2 Butts	"	1 Tun,	"	tun.

Note.—By a law of Massachusetts, the barrel for cider and beer shall contain 32 gallons, but in some other States it is of different capacity. The Ale Gallon contains 282 cubic or solid inches.

TABLE OF TIME.

60 Seconds (*sec.*)	make	1 Minute,	marked	m.
60 Minutes	"	1 Hour,	"	hr.
24 Hours	"	1 Day,	"	d.
7 Days	"	1 Week,	"	w.
4 Weeks	"	1 Month,	"	mo.
12 Calendar months	"	1 Year,	"	yr.
52 Weeks	"	1 Year,	"	yr.
365 Days	"	1 Common Year,	"	yr.
366 Days	"	1 Leap Year,	"	yr.
100 Years	"	1 Century,	"	C.

The following table exhibits the divisions of the year, the names of the months, and the number of days in each:

Winter.	1st month,	January,	has	31 days,
	2d "	February,	"	28, in leap year 29.
Spring.	3d "	March,	"	31 days.
	4th "	April,	"	30 "
	5th "	May,	"	31 "
Summer.	6th "	June,	"	30 "
	7th "	July,	"	31 "
	8th "	August,	"	31 "
Autumn.	9th "	September,	"	30 "
	10th "	October,	"	31 "
	11th "	November,	"	30 "
Winter.	12th "	December,	"	31 "

The following lines will help to remember the number of days in each month:

> "Thirty days hath September,
> April, June, and November;
> All the rest have thirty-one,
> Except February alone,
> Which hath but twenty-eight, in fine,
> Till leap year gives it twenty-nine."

TABLE OF CIRCULAR MOTION.

6o Seconds, or 60 '	make	1 Prime minute,	marked	
60 Minutes	"	1 Degree,	"	♥.
30 Degrees	"	1 Sign,	"	s.
12 Signs, or 860 Degrees, the whole great circle of the zodiac.				

MISCELLANEOUS TABLE.

A gallon of train oil	weighs	7½	pounds.
A stone of butcher's meat	"	8	"
A gallon of molasses	"	11	"
A stone of iron	":	14	"
A tod	"	28	"
A firkin of butter	"	56	"
A firkin of soap	"	94	"
A quintal of fish	"	100	"
A weigh	"	182	"
A sack	"	364	"
A puncheon of foreign prunes	"	1120	"
A last	"	4868	"
A fother of lead	"	19½	cwt.
A bbl. of flour	"	196	pounds.
A " anchovies	"	30	"
A " raisins	"	112	"
A " pork or beef	"	200	"
A ". soap	"	256	"
A " shad or salmon in Connecticut or New York	"	200	"
A " fish in Massachusetts	is	30	gallons.
A " cider and beer	is	82	"
A " herrings in England	is	82	" /
A " salmon or eels do.	is	42	"
8 bushels of salt, measured on board the vessel,	is	1	hogshead.
7½ do. measured on shore,	is	1	"
5 hoops	make	1	cast.
40 casts	"	1	hundred.
10 hundred	"	1	thousand.
12 units, or things,	"	1	dozen.
12 dozen	"	1	gross.
144 dozen	"	1	great gross.
24 sheets of paper	"	1	quire.
20 quires	"	1	ream.
56 pounds	"	1	bush. corn.
60 pounds	"	1	bush. wheat

BOOKS.

A sheet folded in 2 leaves	is called a *Folio.*
A sheet folded in 4 leaves	is called a *Quarto,* or 4to.
A sheet folded in 8 leaves	is called an *Octavo,* or 8vo.
A sheet folded in 12 leaves	is called a *Duodecimo,* or 12mo.
A sheet folded in 18 leaves	is called an 18mo.
A sheet folded in 24 leaves	is called a 24mo.

POPULAR QUESTION BOOK

FOR SUNDAY SCHOOLS.

Primary Bible Questions,

FOR

YOUNG CHILDREN,

BY S. SOOT.

THIRD EDITION, REVISED, ENLARGED AND IMPROVED.

The Publisher takes pleasure in stating that this edition of PRIMARY BIBLE QUESTIONS has been received with increased favor, and that it has been introduced, and is now in use, in many of the largest Sabbath Schools without denominational restrictions, in Georgia and Alabama.

To meet the demand for such books, a large edition has been issued, and it is now offered to schools at $3,00 per dozen.— Single copies 30 cents.

SUNDAY SCHOOL TESTAMENTS

At 10 cents per copy can be had by addressing

J. J. TOON, Publisher,
And Proprietor Franklin Printing House,
Atlanta, Georgia.